Bibliographic information published by the German National Library:

The German National Library lists this publication in the National Bibliography;
detailed bibliographic data are available on the Internet at http://dnb.dnb.de .

Imprint:

Copyright © 2015 GRIN Verlag, Open Publishing GmbH
Print and binding: Books on Demand GmbH, Norderstedt Germany
ISBN: 9783668270985

This book at GRIN:

http://www.grin.com/en/e-book/334979/optimized-multi-modes-mimo-positive-
position-feedback-active-vibration

Zhonghui Wu

Optimized Multi-modes MIMO Positive Position Feedback Active Vibration for Plate Structure

GRIN Publishing

Contents

Optimized Multi-modes Multi-Input-Multi-Output Positive Position Feedback Active Vibration for Plate Structure

Zhonghui WU

School of Computer Science, Engineering and Mathematics, Faculty of Science and Engineering, Flinders University, Australia

---

ABSTRACT

Active vibration control (AVC) methodology is presented by the author in this paper using bonded three self-sensing magnetic transducers for a flexible plate structure. Multi-modes multi-input-multi-output (MIMO) positive position feedback (PPF) controller is tested and verified for vibration suppression through simulation and experiment implement. Based on genetic algorithm (GA) searching, optimal parameters of the controllers can be obtained according to the minimization criterion which is the solution to the H_∞ norm of the whole closed-loop system.

Keywords

Multi-Input-Multi-Output, Positive Position Feedback, Genetic Algorithm, Vibration Control, Optimization

---

1. Introduction

In order to decrease the cross-sectional dimensions of the structures, improve dynamic performance and operating efficiency, lightweight products and materials were employed by many designers. However, when the structures become flexible, the harmful effects of unwanted vibration can be seen from structures, they are becoming more susceptible.

The problem is extremely worse when they operate at or near their natural frequencies or when they are excited by disturbances that coincide with their natural frequencies [1].

Modal control has become the best choice for vibration control engineers to suppress the vibration of flexible structures for many years. In general, modal analysis and control refer to the procedure of decomposing the dynamic equations of mechanical system into modal coordinates and designing the control system in this modal coordinate system [2]. It extracts the interested mode signal from the structural response. The engineer can design the controller in the modal domain and control a single degree-of-freedom oscillator in the similar way [3]. There are three main modal control methods that can be found in the literature for controlling multi-modes vibration in flexible structures: independent modal space control (IMSC), resonant control and positive position feedback (PPF) control.

Meirovitch provided the independent modal space control (IMSC). IMSC can design the controller for each single mode. The controller can be implemented independently, in that way there will be little spillover to the residual modes [4-5]. But the problem is obvious: it needs lots of sensors/ actuators as the number of modes, which need to be controlled, and it can control a limited number of modes. Furthermore, the control system is not robust to uncertainties such as parameter fluctuation. In order to solve these problems, engineers had tried to apply the robust control techniques to this area in the past two decades [6].

According to the investigation of the resonant characteristic of the flexible structures, Moheimani raised resonant control method [7-12]. The designer chose a high gain controller at the natural frequency of the flexible structure. The controller would roll off quickly away from the natural frequency in order to avoid spillover. It is also described as having a decentralized characteristic from a modal control perspective [13], thereby making it possible to treat each mode of the system in isolation. One of limited performances for resonant controllers is restricted increase damping to the structure [14].

After compared with other methods, Goh and Caughey provided Positive position feedback (PPF) [15,16], the stability was proved by Fanson J. L. [17,18]. Compared with other methods, PPF controller has several significant advantages [19]. It has been proved by many researchers that the

PPF is a reliable vibration control strategy to suppress the vibration of flexible systems with smart materials [15-17].

After that, lots of people did simulations and experiments using the PPF theory for active vibration [20-33]. Based on the performance has already achieved, some researchers modify the structure of PPF [34-40]. Expect the normal advantages, some people also showed the robust ability of the PPF controller [41-44].

During the PPF controller design, in order to achieve good performance, some designer used Genetic Algorithms (GA) to choose the placement of sensor and actuator. [45] applied GA to find efficient location of sensor and actuator of a cantilevered composite plate, vibration reduction for the first three modes has been achieved using the coupled PPF. The designer has proved that the whole system is robust to parameter variations. [46] presented a GA method of optimal placement for the cantilever plate. According to adding PD and PPF together, the proposed control method by can suppress the vibration decay process and the smaller amplitude vibration effectively, which has been approved and can be seen from simulations and experimental results.

In order to achieve better performance, based the complicated coupled relations between controllers, some designers provided MIMO PPF controller.

For beam structure: MIMO PPF controller has been verified by [47,14] on a cantilevered beam according to experimental implementation. The author also adopted pole placement and H_∞ optimization method. [48] designed MIMO PPF controller on a flexible manipulator and through the GA method to find controller parameters to optimize H_∞ result. [62] designed and compared HMPPF and HMVPF controller, which is followed [47,14] method to control multi-modes of the beam structure. MPVF controller was designed and provided by [64] for the vibration suppression of the beam.

For grid structure: [49-51] proposed the use of the GA method for tuning MIMO PPF controller for grid structure. Based on the block-inverse technique, [52] provided MIMO PPF controller to suppress the vibration of the grid structure.

For plate structure: according to pseudo-inverse technique, [53] designed and proposed MIMO PPF controller. A nonlinear MIMO PPF control method is presented by [56], both high and low amplitude vibration suppression of the flexible cantilever plate was achieved. [57] provided MIMO PPF and PD combined controller to control the decoupled bending and torsional modes of the plate. MIMO PPF and MIMO SISO were compared by [63] to suppress vibration of sandwich plate.

For shell structure, using block-inverse technique. [54] studied and verified MIMO PPF controller for the first two vibration modes.

For switched reluctance machine, based on the studying of PPF theory [55] gave a decentralized MIMO experimental compensation method.

In this paper, the author will use the MIMO PPF controller based H_∞ optimization through GA searching for multi-modes vibration control of a flexible plate structure. According to the author's understanding, this effort is the first time to apply to plate structure, similar methods can be only found on beam structure.

2. Model of Flexible Plate Structure

2.1 Experimental Model

The schematic of the whole system, which is designed as an experimental plant to test the controller, is presented in Fig. 2.1. A uniform AL6061-T6 plate mounts with screw on the MDF board using three electromagnetic transducers (anticlockwise number 1, 2, 3). Another electromagnetic transducer is mounted on the MDF board with screw as a disturbance noise shaker. The MDF board is placed on the table with four rubber legs, which are screwed to the MDF board.

Figure 2.1   A thin plate in transverse vibration

The mechanical model of the transducer, which is used in the experimental plant, is presented in Figure 2.2a. The structure of the transducer is similar to a loudspeaker, which also consists of a coil, produces a magnetic field when a current is fed through [58]. The magnet of the core, which is mounted inside the coil, will be affected by the vibration of the plate [58]. The electric circuit model, which is used for this transducer, is shown in Figure 2.2b.

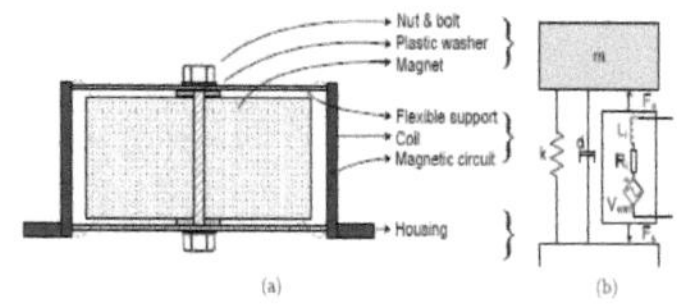

Figure 2.2   Cross section of transducer [58]

It is supposed that the transducers are mechanically identical. The measured mechanical parameters of the transducers can be seen in Table 2.1.

Parameter	Symbol	Value
Spring constant	k	19037 N/m
Core mass	m	0.323 kg
Damping ratio	ζ	$3.39 \cdot 10^{-2}$
Damping Coefficient	d	5.32 Ns/m
Damped resonance frequency	ω_d	38.63 Hz
Resonance frequency	ω_c	38.65 Hz

Table 2.1: Mechanical parameters of transducer

Applying a sweep signal to experimental model, then the natural frequencies of the whole plant are obtained. According to observing the frequency of the signal, the amplitude of the vibration, which is needed to test, can be measured and shown on the oscilloscope by accelerometers.

Experiments show that the dominant modes for all the experimental models are the first four modes: 27.1 Hz, 34.4 Hz, 40.5 Hz, 49.2 Hz.

2.2 Analytical Model

This analytical model concludes a thin plate and four supporting transducers. These main assumptions are based on theory from [59]: a thin plate with dimensions (a * b * h). The Young's modulus of elasticity, density and Poisson's ratio of the plate are denoted by E, ρ, and ν respectively. The deflection is in X and Y directions.

PDE of a thin plate under transverse vibration [59]:

$$\rho(x, y)\, h \frac{\partial^2 w}{\partial t^2} + D\nabla^4 w(t, x, y) = \frac{\partial^2 M_x}{\partial x^2} + \frac{\partial^2 M_y}{\partial y^2} + p_z(t, x, y) \tag{2.1}$$

where

$$\nabla^4 w = \frac{\partial^4 w}{\partial x^4} + 2\frac{\partial^4 w}{\partial x^2\, \partial y^2} + \frac{\partial^4 w}{\partial y^4} \tag{2.2}$$

2.3 Modal Analysis

Consider the typical PDE for flexible structures [1]:

$$\mathcal{L}[w(x, y, t)] + C\left[\frac{\partial w(x,y,t)}{\partial t}\right] + \mathcal{M}\left[\frac{\partial^2 w(x, y,t)}{\partial t^2}\right] = f(x, y, t) \tag{2.3}$$

$\mathcal{L}$ and $\mathcal{M}$ are linear homogeneous differential operators of order 2p and 2q respectively and q $\leq$ p. Here, x, y is the spatial coordinate, which is defined over a domain $\mathcal{D}$. The general arbitrary input is denoted by f, which is distributed over $\mathcal{D}$. The boundary conditions corresponding to PDE (2.103) can be expressed as

$$\mathcal{B}i\,[w(x,y,t))] = 0, \quad i = 1, 2, ..., p \quad (2.4)$$

where $\mathcal{B}i$ is a linear homogeneous differential operator of order less than or equal to $2p - 1$.

The transfer function of the system can be derived according to [1]:

$$G\,(s, x, y) = \sum_{m=1}^{\infty} \frac{\phi_{mn}(x,y,t)P_m}{s^2 + 2\xi_m \omega_m s + \omega_m^2} \quad (2.5)$$

Because there is an infinite number of modes, so the Equation (2.5) is an infinite-dimensional transfer function. This is a general solution of PDE (2.3). The solution for a particular structure is then solved by finding the eigenfunction $(\phi_{mn}(x,y,t))$, the natural frequency (ω_m), and the structural damping (ξ_m).

In order to solve the partial differential equation (2.1) for uniform thin plates with four supporting transducers under transverse vibration, the modal analysis methodology will be applied to find the solution. The solution is assumed to be: $w(t, x, y) = \sum_{m=1}^{\infty} \sum_{n=1}^{\infty} \phi_{mn}(x,y,t)\, q_{mn}(x,y,t)$, where $q_{mn}(x,y,t)$ and $\phi_{mn}(x,y,t)$ are the generalized coordinate and the eigenfunction, respectively. Based on the orthogonality properties of the eigenfunctions $\phi_{mn}(x,y,t)$, PDE of thin plate with four supporting transducers under transverse vibration can be solved independently for each mode [59].

Based on the modal analysis solution, the transfer function from the actuator voltages, $V_a(s) = [V_{a1} ... V_{aI}]^T$, to the plate deflection $w(x, y, s)$ can be written as

$$G(s,x,y) = \sum_{m=1}^{\infty} \sum_{n=1}^{\infty} \frac{\phi_{mn}(x,y,t)\bar{\Psi}_{mn}^T}{s^2 + 2\xi_{mn}\omega_{mn}s + \omega_{mn}^2} \quad (2.6)$$

where $(x, y) \in \mathbb{R}$, $\mathbb{R} = \{(x,y) \mid 0 \leq x \leq L_x, 0 \leq y \leq L_y\}$ and ξ_{mn} is the damping ratio associated with the mode (m, n). Furthermore, $\bar{\Psi}_{mn} = [\Psi_{mn1} ... \Psi_{mnI}]^T$ will be affected by the properties of the plate, actuator place and the eigenfunctions $\phi_{mn}(x,y,t)$.

2.4 Numerical Model

Using software tool ANSYS, numerical model is designed and shown in Fig 2.3

2.5 Simulation Model of The Plate

Due to the MIMO properties, any one of the outputs on the top plate is induced by superposing all the inputs and their related dynamics. Hence, the dynamics of the whole system can be expressed by output-input

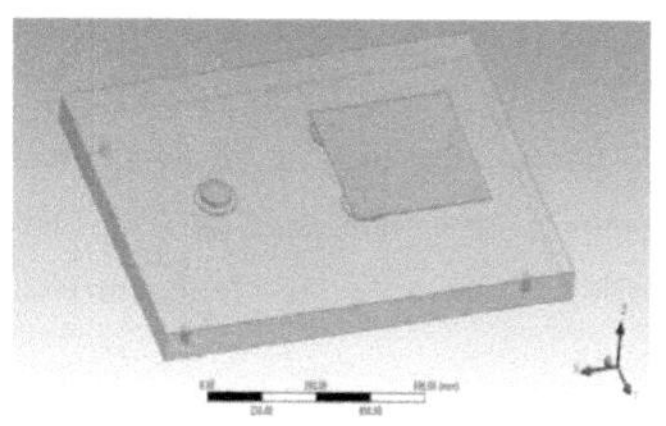

Figure 2.3   ANSYS Model

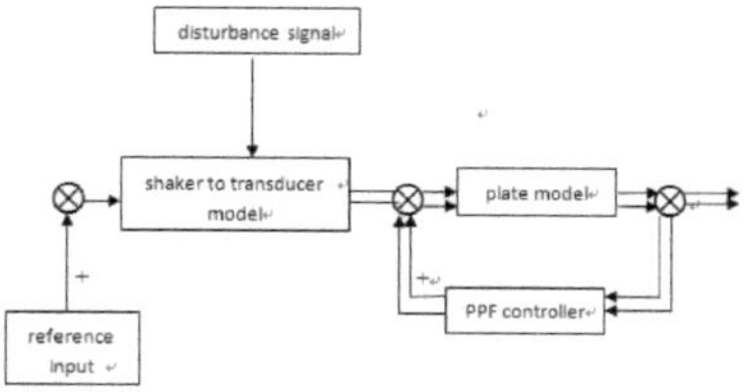

disturbance signal
shaker to transducer model
plate model
PPF controller
reference input

Figure 2.4   Block diagram of PPF control

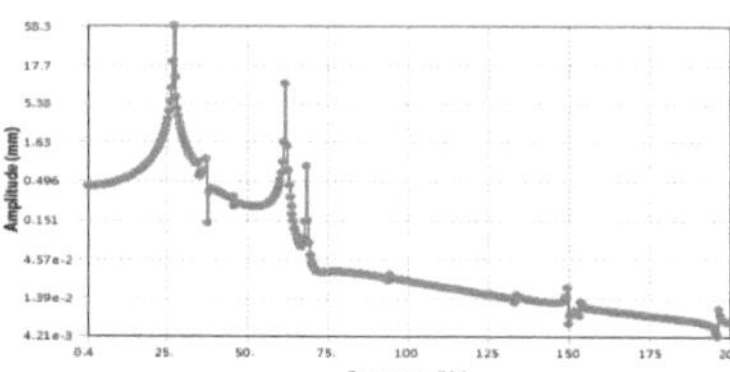

Figure 2.5 Amplitude Response

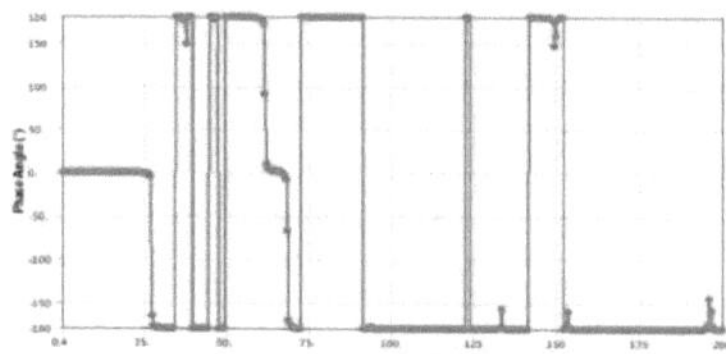

Figure 2.6 Phase Response

open loop system superposition full description.

$$\begin{bmatrix} Y_1(s) \\ Y_2(s) \\ Y_3(s) \end{bmatrix} = \begin{bmatrix} T_{11}(s) & T_{12}(s) & T_{13}(s) \\ T_{21}(s) & T_{22}(s) & T_{23}(s) \\ T_{31}(s) & T_{32}(s) & T_{33}(s) \end{bmatrix} * \begin{bmatrix} U_1(s) \\ U_2(s) \\ U_3(s) \end{bmatrix}$$

(2.7)

where

$$\begin{bmatrix} U_1(s) \\ U_2(s) \\ U_3(s) \end{bmatrix} = \begin{bmatrix} D_1(s) \\ D_2(s) \\ D_3(s) \end{bmatrix} * U_d(s) \qquad (2.8)$$

U_i – input through the transducer i to the top plate, i = 1, 2, 3; U_d – input to shaker; Y_i - output at the position on the top plate, i = 1, 2, 3; D_i - dynamic between the shaker and transducer; T_{mn} - dynamic between transducer m and n, m = 1, 2, 3, n = 1, 2, 3.

3. Multi-modes MIMO PPF Controller

3.1 PPF Controller

Positive position feedback (PPF) controller was first proposed, applied and verified using experimental implement by Caughey and Goh in 1982. Its simplicity and robustness has attracted many researchers to adopt and apply in flexible structural vibration controlling [16]. According to observing, the researchers find that PPF is different from other control laws, the PPF controller is insensitive to uncertain natural damping ratios of the structure [16]. The measurement of position is positive, then it is fed into the compensator. And at the same time, the position signal from the compensator, which is fed back to the structure later, is also positive [18]. This character makes the PPF controller quite fit for collocated actuator/sensor pairs.

Equations 3.1 and 3.2 show the structure and compensator equations in the scalar case [16]:

Structure: $\ddot{\varepsilon} + 2\xi\omega\dot{\varepsilon} + \omega^2\varepsilon = g\omega^2\eta$ (3.1)

Compensator: $\ddot{\eta}+2\xi_f\omega_f\dot{\eta}+\omega_f^2\eta =\omega_f^2\varepsilon$ (3.2)

where g is the scalar gain (positive), ε is the modal coordinate (structural), η is the filter coordinate (electrical), ω are the structural frequencies, ω_f are the filter frequencies, ζ are the structural damping ratios and $2\xi_f$ are the filter damping ratios. This non-dynamic stability criterion is characteristic of the positive position feedback system.

PPF compensator can be seen in Equation (3.3) as the second-order transfer function. The transfer function form will be used in this text for deriving the properties of the control system [16].

$$G_{ppf}(s) = \frac{g\omega_f^2}{s^2+ 2\xi_f\omega_f s+ \omega_f^2} \qquad (3.3)$$

Another feature of the compensator equation, its second-order low-pass characteristic, led to the terminology PPF filter. In effect, a PPF filter behaves much like an electronic vibration absorber for the structure, which is different from a mechanical vibration absorber [1].

The feedback controllers for multivariate resonant systems can be written as [60,1]:

$$G(s)= \sum_{i=1}^{M} \frac{\psi_i\psi_i'}{s^2+ 2\xi_i\omega_i s+ \omega_i^2} \qquad (3.4)$$

where ψ_i is an mx1 vector, and M $\rightarrow \infty$. In practice, we hope that the integer is finite. This is because that only including a very large number of modes, equation (3.4) can be sufficiently described the excitation of elastic structure.

An important character of PPF controllers is that only requires one second-order term to repress one vibration mode. So the designer can choose the modes, which are needed to be controlled in a specific bandwidth [60].

In order to deliver stability conditions for this control loop, the series in (3.4) is truncated by keeping the first N modes (N<M). These N modes are in the bandwidth of interest [60,61]. A feed-through term is added to the truncated model to remedy the effect of truncated modes [60,61]. The series in (3.4) are approximated as [60,61]:

$$G^N (s)= \sum_{i=1}^{N} \frac{\psi_i\psi_i'}{s^2+ 2\xi_i\omega_i s+ \omega_i^2} + D \qquad (3.5)$$

3.2. Multi-modes MIMO PPF Controller Parameter Selection

Based the dynamics model between shaker and plate, the results of multi-modes MIMO PPF controller optimal parameters, are found through GA searching minimum value of H_∞ norm using MATLAB toolbox, shown in Fig3.1.

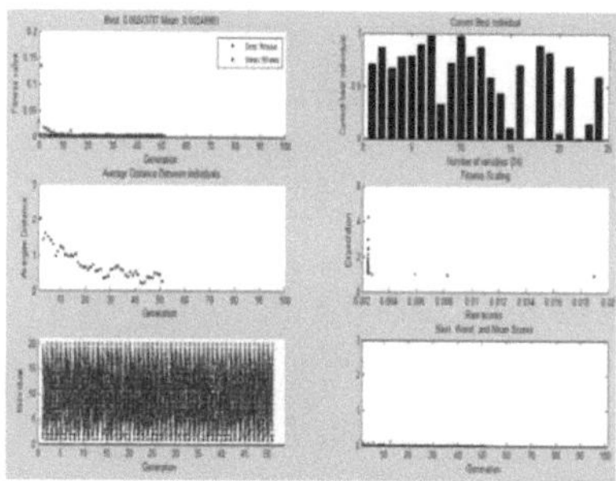

Figure 3.1 optimal parameter results through GA

4. Simulation

The MIMO PPF controller is used to control the first four vibration modes of the simulation plate structure model using MATLAB.

Firstly, simulation plate structure model and MIMO PPF controller are connected as feedback closed-loop system.

Secondly, dynamics correction model from the shaker to the transducer is connected to feedback closed-loop system as in Fig2.4.

Thirdly, the parameters of the controller are chosen from the GA searching result.

Fourthly, sweep signal, which contains resonant frequencies of the first four modes of the simulation plate structure model, is applied to the shaker.

Fifthly, Open-loop dynamics, which is between shaker to plate without the controller, is compared to closed-loop dynamics, which is between shaker to plate with the controller.

Time domain and frequency domain cases are designed to test the ability of the MIMO PPF controller to attenuate multi-modes vibration when the controller centre frequencies are equal to the resonant frequencies of the plate.

5. Experiment

The MIMO PPF controller is used to control the first four vibration modes of the experimental plate structure model.

Firstly, the MIMO PPF controller is formed in MATLAB Simulink.

Secondly, sweep signals, which contain resonant frequencies of the first four modes of the simulation plate structure model, are applied to the shaker one by one.

Thirdly, Open-loop dynamics, which is between shaker to plate without the controller, is compared to closed-loop dynamics, which is between shaker to plate with the controller.

Time domain cases are designed to test the ability of the MIMO PPF controller to attenuate multi-modes vibration when the controller centre frequencies are equal to the resonant frequencies of the plate.

A technique called 'self-sensing' was used to measure the back-emf voltage without using any additional sensors. The transducer is performed collocated as it is used for actuation and sensing at the same time [58].

DSpace DS1103 Controller Board was used to interface with the system to be able to acquire the back-emf voltage and to output the reference transducer voltage. 16 16-bit   analog-to-digital (AD) converters, 4 12-bit

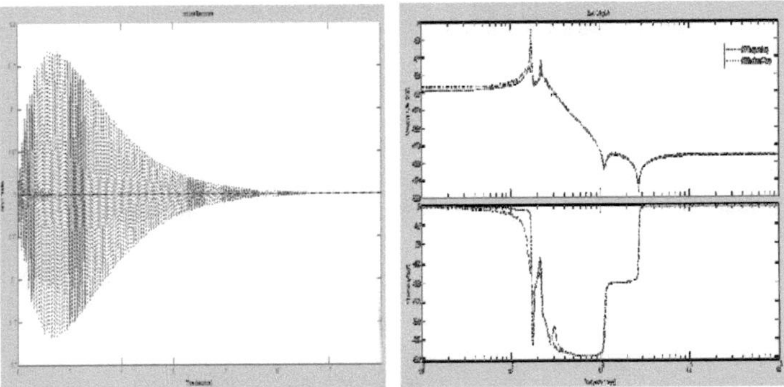

Figure 4.1 open-loop, closed-loop simulation result at transducer 1

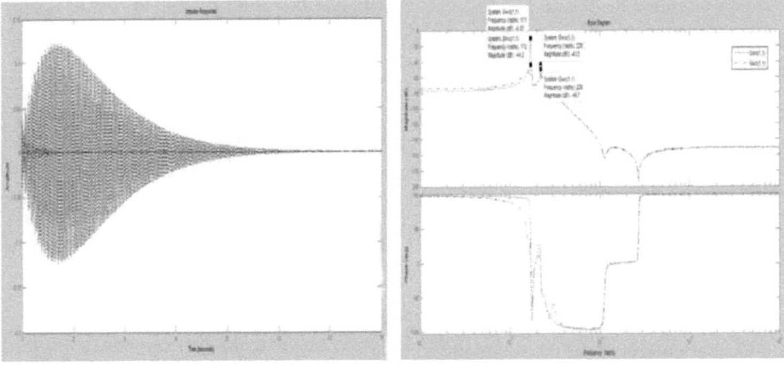

Figure 4.2 open-loop, closed-loop simulation result at transducer 2

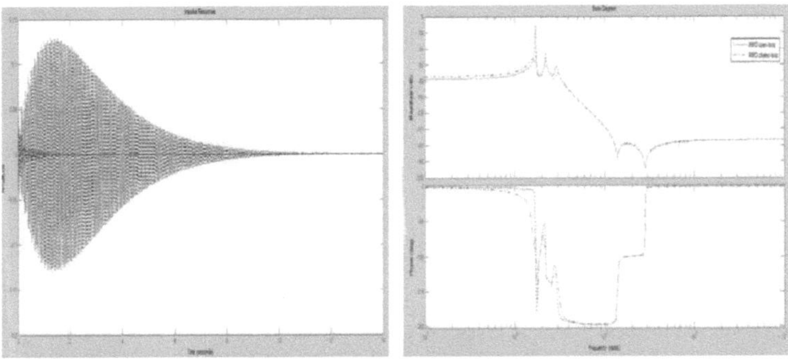

Figure 4.3 open-loop, closed-loop simulation result at transducer 3

AD converters and 8 14-bit digital-to-analog (DA) converters are provided. Onboard DSP is used for the signal processing and acquisition the output of the various signals via its AD and DA converters when running a simulation externally in MATLAB SIMULINK [58].

The disturbance transducer was powered by a 50 W Jaycar amplifier kit in order to induce vibrations in the system. A sinusoidal signal, which frequency was chosen from 20 to 50 Hz, was used as the input signal, sampling time is 0.1s.

Open-loop dynamics, which is between shaker to plate without the controller, is compared to closed-loop dynamics, which is between shaker to plate with the controller. Frequency domain and time domain cases are designed to test the ability of the MIMO controller to attenuate multi-modes vibration when the controller centre frequencies are equal to the resonant frequencies of the plate.

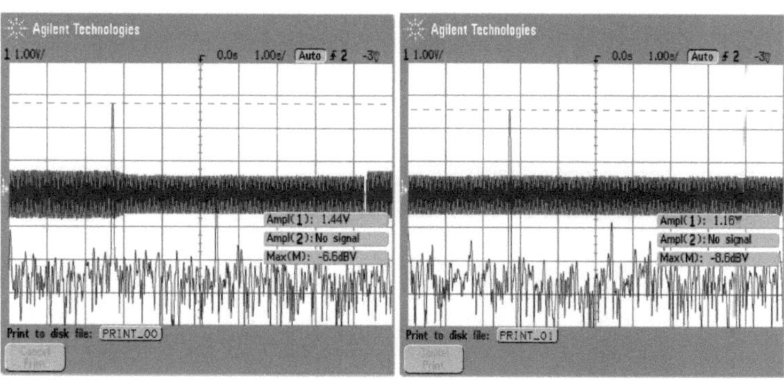

Figure 5.1 transducer 1 first mode 27.1 Hz before and after control

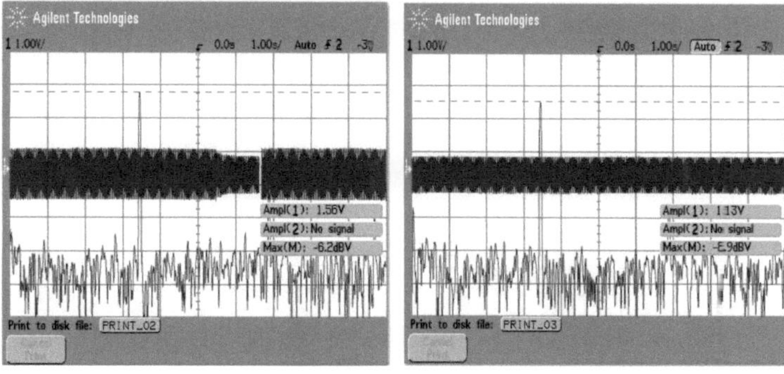

Figure 5.2 transducer 1 second mode 34.4 Hz after control

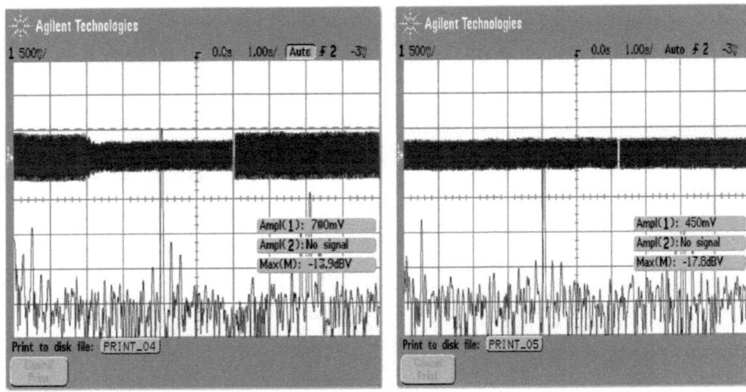

Figure 5.3 transducer 1 third mode 40.5 Hz before and after control

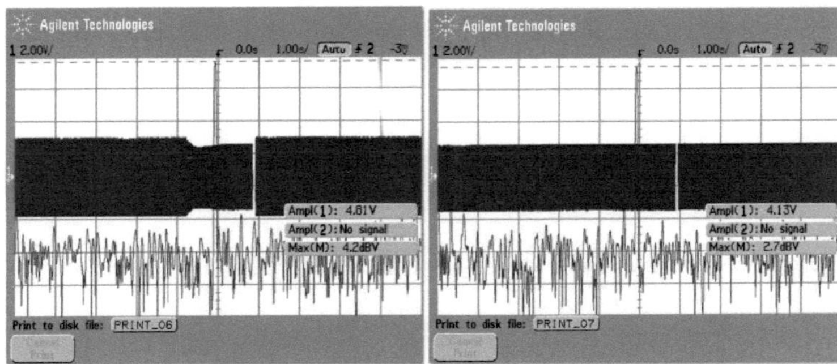

Figure 5.4 transducer 1 forth mode 49.2 Hz before and after control

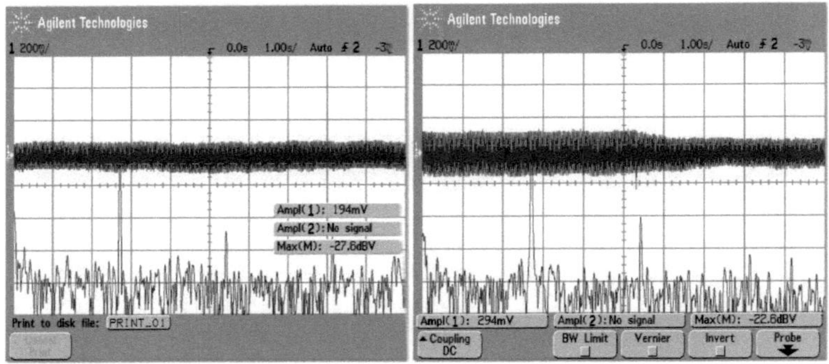

Figure 5.5 transducer 2 first mode 27.1Hz before and after control

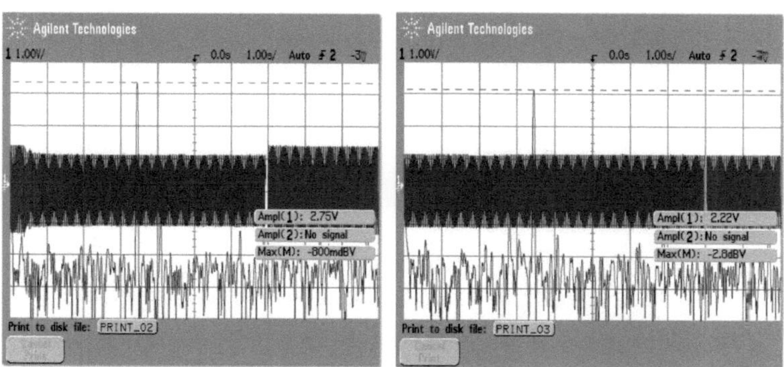

Figure 5.6 transducer 2 second mode 34.4 Hz before and after control

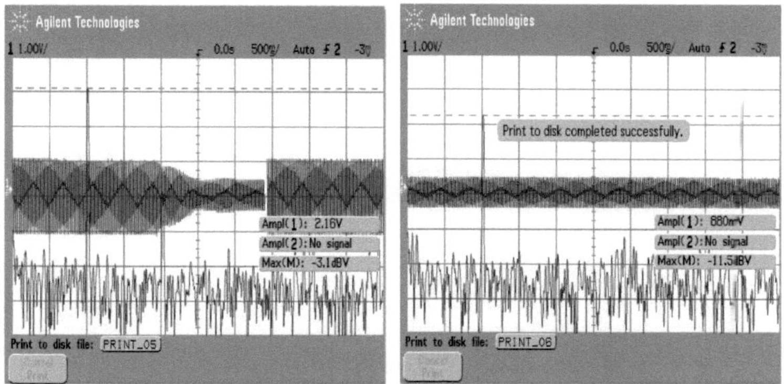

Figure 5.7 transducer 2 third mode 40.5 Hz before and after control

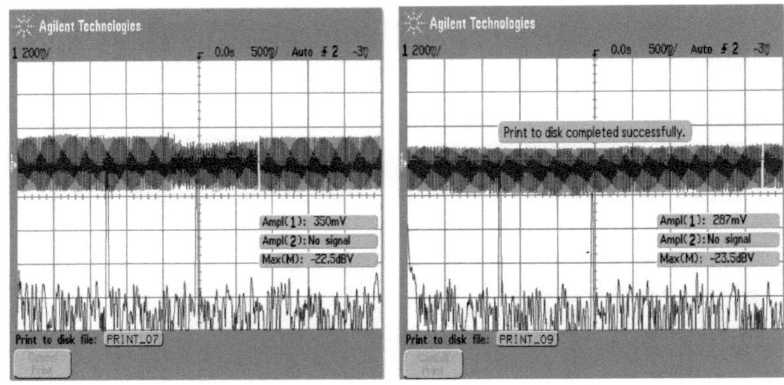

Figure 5.8 transducer 2 forth mode 49.2 Hz before and after control

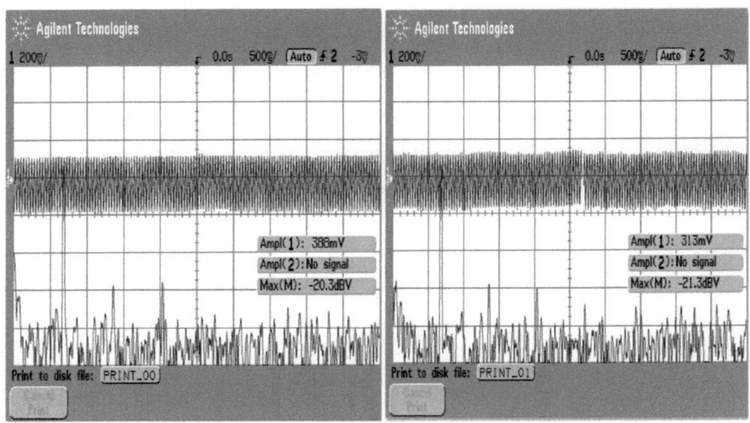

Figure 5.9 transducer 3 first mode 27.1Hz before after control

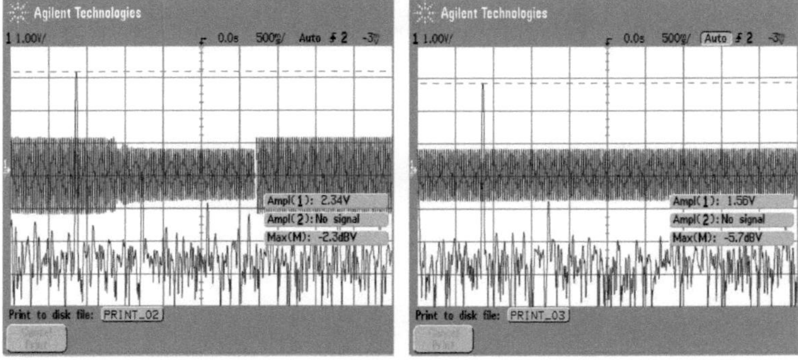

Figure 5.10 transducer 3 second mode 34.4 Hz before and after control

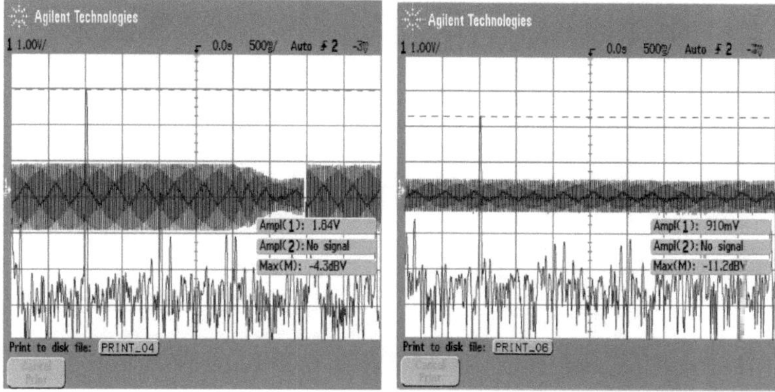

Figure 5.11 transducer 3 third mode 40.5 Hz before and after control

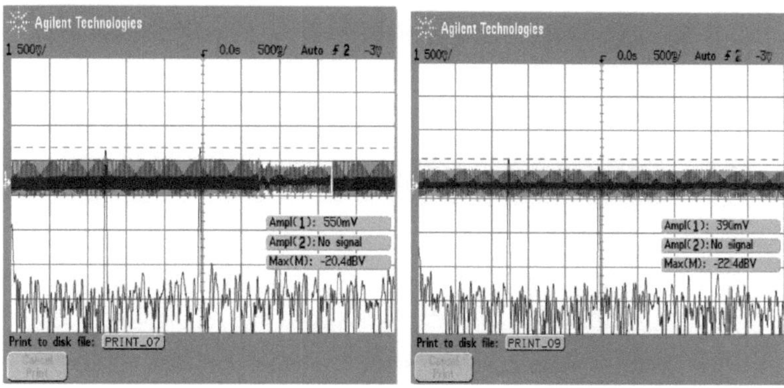

Figure 5.12 transducer 3 forth mode 49.2 Hz before and after control

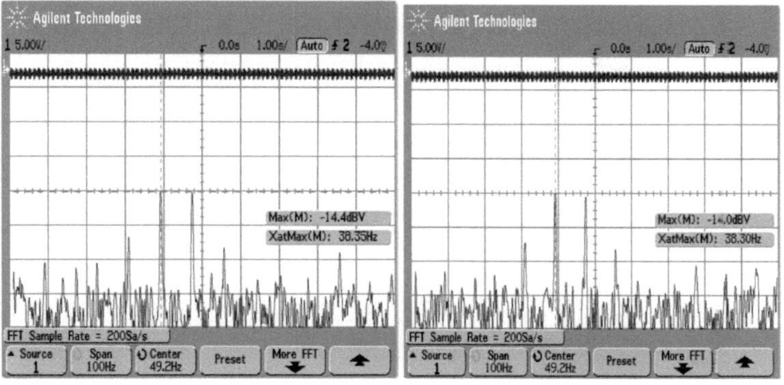

Figure 5.13 sweep signal at transducer 1 before and after control

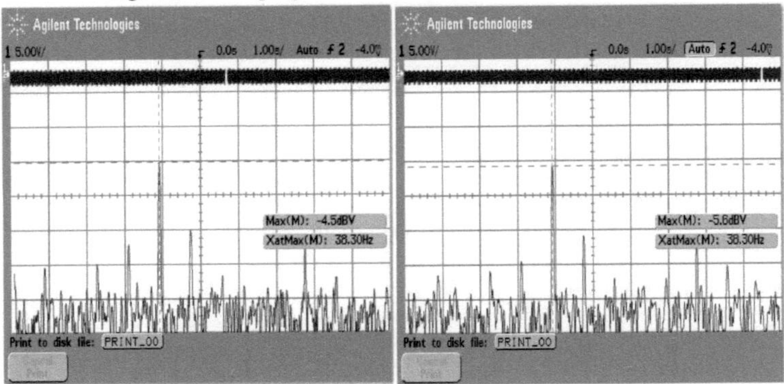

Figure 5.14 sweep signal at transducer 2 before and after control

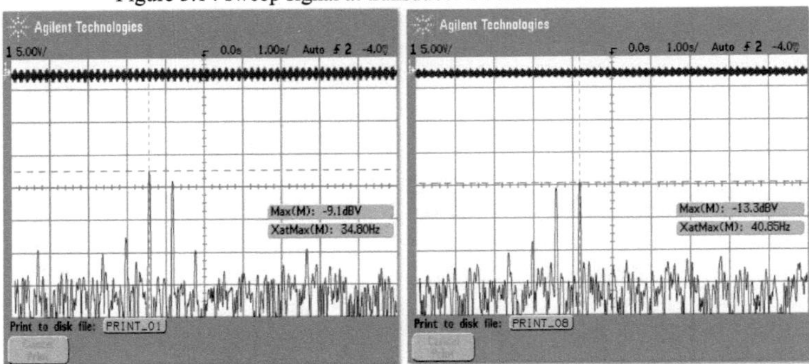

Figure 5.15 sweep signal at transducer 3 before and after control

			frequency (Hz)			
			27.1	34.4	40.5	49.2
transducer 1	Max Peak	before	-6.6	-6.2	-13.9	4.2
	(dB)	after	-8.6	-8.9	-17.8	2.7
	reduce(dB)		2	2.7	3.9	1.5
	reduce rate (%)		**19.444%**	**27.64%**	**35.714%**	**14.137%**
transducer 2	Max Peak	before	-22.6	-0.8	-3.1	-22.5
	(dB)	after	-27.6	-2.8	-11.5	-23.5
	Reduce (dB)		5	2	7.6	1
	reduce rate(%)		34.01%	19.27%	59.26%	18%
transducer 3	Max Peak	before	-20.3	-2.3	-4.3	-20.4
	(dB)	after	-21.3	-5.7	-11.2	-22.4

	Reduce (dB)	1	3.4	6.9	2
	reduce rate(%)	**19.33%**	**33.33%**	**50.543%**	**29.091%**

Table 5.1   multi-modes MIMO vibration control experimental results

6. Summary

According to simulation and experiment result, the provided methodology H_∞ optimization MIMO positive position feedback based on genetic algorithm can be successfully used for the multi-modes active vibration suppression of plate structure. At

MIMO positive position feedback controller, which can be applied to multi-modes vibration suppression of a large class of

the same time, we achieve good vibration suppression results at first four modes of the plate. Compared to other methods, there is no noise happened at the beginning of turning on the controller and spillover. Future work is preparing for the implementation of adaptive active vibration control methodology using optimized

flexible structures with varying and unknown parameters or loading conditions.

Acknowledgement

S.S. Groothuis and R.J. Roesthuis contributed to design the experiment plant and ANASYS simulation model in 2010.

Siyang Yu contributed to modify the ANSYS simulation model in 2012.

References

[1] Tjahyadi, H. (2007) Adaptive multi mode vibration control of dynamically loaded flexible structures (Doctoral dissertation, Flinders University)

[2] Li, S. (2011). Active modal control simulation of vibro-acoustic response of a fluid-loaded plate. Journal of Sound and Vibration, 330(23), 5545-5557

[3] Kim, S. M., & Oh, J. E. (2013). A modal filter approach to non-collocated vibration control of structures. Journal of Sound and Vibration

[4] Meirovitch, L., Van Landingham, H. F., & Öz, H. (1977). Control of spinning flexible spacecraft by modal synthesis. Acta Astronautica, 4(9), 985-1010

[5] Meirovitch, L., Baruh, H., & OZ, H. (1983). A comparison of control techniques for large flexible systems. Journal of Guidance, Control, and Dynamics, 6(4), 302-310

[6] Iwamoto, H., Tanaka, N., & Hill, S. G. (2012). Feedback control of wave propagation in a rectangular panel, part 2: Experimental realization using clustered velocity and displacement feedback. Mechanical Systems and Signal Processing

[7] Moheimani, S. O. R., Fleming, A. J., & Behrens, S. (2001). Highly resonant controller for multimode piezoelectric shunt damping. Electronics Letters, 37(25), 1505-1506

[8] Pota, H. R., Moheimani, S. R., & Smith, M. (2002). Resonant controllers for smart structures. Smart Materials and Structures, 11(1), 1

[9] Aphale, S. S., Fleming, A. J., & Moheimani, S. R. (2007). Integral resonant control of collocated smart structures. Smart Materials and Structures, 16(2), 439

[10] Pereira, E., Moheimani, S. O. R., & Aphale, S. S. (2008). Analog implementation of an integral resonant control scheme. Smart Materials and Structures, 17(6), 067001.

[11] Mahmood, I. A., Moheimani, S. R., & Bhikkaji, B. (2008). Precise tip positioning of a flexible manipulator using resonant control. Mechatronics, IEEE/ASME Transactions on, 13(2), 180-186

[12] Pereira, E., Aphale, S. S., Feliu, V., & Moheimani, S. R. (2011). Integral resonant control for vibration damping and precise tip-positioning of a single-link flexible manipulator. Mechatronics, IEEE/ASME Transactions on, 16(2), 232-240

[13] Halim, D., & Moheimani, S. R. (2001). Spatial resonant control of flexible structures-application to a piezoelectric laminate beam. Control Systems Technology, IEEE Transactions on, 9(1), 37-53

[14] Moheimani, S. R., Vautier, B. J., & Bhikkaji, B. (2006). Experimental implementation of extended multivariable PPF control on an active structure. Control Systems Technology, IEEE Transactions on, 14(3), 443-455

[15] Goh, C. J. (1983). Analysis and control of quasi distributed parameter systems (Doctoral dissertation, California Institute of Technology)

[16] Goh, C. J., & Caughey, T. K. (1985). On the stability problem caused by finite actuator dynamics in the collocated control of large space structures. International Journal of Control, 41(3), 787-802

[17] Fanson J L, An experimental investigation of vibration suppression in large space structures using positive position feedback, PhD Thesis California Institute of Technology,November,1986

[18] J. L. Fanson, and T. K. Caughey, "Positive Position Feedback Control for Large Space Structures," AIAA Journal, vol. 28, no. 4, pp.717-724, 1990

[19] Song G, Schmidt S P and Agrawal B N, Experimental robustness study of positive position feedback control for active vibration suppression, J. Guidance, VOL. 25, NO. 1: ENGINEERING NOTES, 2001

[20] J. L. Fanson, and T. K. Caughey, "Positive Position Feedback Control for Large Space Structures," AIAA Journal, vol. 28, no. 4, pp.717-724, 1990

[21] Song G, Schmidt S P and Agrawal B N, Experimental robustness study of positive position feedback control for active vibration suppression, J. Guidance, VOL. 25, NO. 1: ENGINEERING NOTES, 2001

[22] Baz, A., Poh, S., & Fedor, J. (1989). Independent modal space control with positive position feedback. Dynamics and control of large structures, 553-567

[23] Poh, S., & Baz, A. (1990). Active control of a flexible structure using a modal positive position feedback controller. Journal of Intelligent Material Systems and Structures, 1(3), 273-288.

[24] Goh, C. J., & Lee, T. H. (1991). Adaptive modal parameters identification for collocated position feedback vibration control. International Journal of Control, 53(3), 597-617.

[25] Caughey, T. K. (1995). Dynamic response of structures constructed from smart materials. Smart Materials and Structures, 4(1A), A101

[26] Baz, A., & Poh, S. (1996). Optimal vibration control with modal positive position feedback. Optimal control applications and methods, 17(2), 141-149

[27] Baz, A., & HONG, J. T. (1997). Adaptive control of flexible structures using modal positive position feedback. International journal of adaptive control and signal processing, 11(3), 231-253

[28] Wang, L. (2003, October). Positive position feedback based vibration attenuation for a flexible aerospace structure using multiple piezoelectric actuators. In Digital Avionics Systems Conference, 2003. DASC'03. The 22nd (Vol. 2, pp. 7-C). IEEE

[29] Shan, J., Liu, H. T., & Sun, D. (2005). Slewing and vibration control of a single-link flexible manipulator by positive position feedback (PPF). Mechatronics, 15(4), 487-503

[30] Hong, C., Shin, C., & Jeong, W. (2010). Active vibration control of clamped beams using PPF controllers with piezoceramic actuators. Proc. 20th Int. Congr. on Acoustics (Sydney, Australia, ICA 2010 23–27 August 2010)

[31] Ahmed, B., & Pota, H. R. (2011). Dynamic compensation for control of a rotary wing UAV using positive position feedback. Journal of Intelligent & Robotic Systems, 61(1-4), 43-56

[32] Chuang, K. C., Ma, C. C., & Wu, R. H. (2012). Active suppression of a beam under a moving mass using a pointwise fiber bragg grating displacement sensing system. Ultrasonics, Ferroelectrics and Frequency Control, IEEE Transactions on, 59(10), 2137-2148

[33] Shin, C., Hong, C., & Jeong, W. B. (2012). Active vibration control of clamped beams using positive position feedback controllers with

moment pair. Journal of mechanical science and technology, 26(3), 731-740

[34] Bang, H., & Agrawal, B. N. (1994). A generalized second order compensator design for vibration control of flexible structures. In AIAA/ASME/ASCE/AHS/ASC Structures, Structural Dynamics, and Materials Conference, 35th, Hilton Head, SC (pp. 2438-2448)

[35] Mahmoodi, S. N., Aagaah, M. R., & Ahmadian, M. (2009, June). Active vibration control of aerospace structures using a modified positive position feedback method. In American Control Conference, 2009. ACC'09. (pp. 4115-4120). IEEE

[36] Hu, Q., Xie, L., & Gao, H. (2006, December). A combined positive position feedback and variable structure approach for flexible spacecraft under input nonlinearity. In Control, Automation, Robotics and Vision, 2006. ICARCV'06. 9th International Conference on (pp. 1-6). IEEE

[37] Hu, Q., Xie, L., & Gao, H. (2007). Adaptive variable structure and active vibration reduction for flexible spacecraft under input nonlinearity. Journal of Vibration and Control, 13(11), 1573-1602

[38] Kim, S. M., Wang, S., & Brennan, M. J. (2011). Comparison of negative and positive position feedback control of a flexible structure. Smart Materials and Structures, 20(1), 015011

[39] Bhikkaji, B., Ratnam, M., Fleming, A. J., & Moheimani, S. R. (2007). High-performance control of piezoelectric tube scanners. Control Systems Technology, IEEE Transactions on, 15(5), 853-866

[40] Chen, L., He, F., & Sammut, K. (2002). Nonlinear modal positive position feedback for vibration control in distributed parameter systems. In Annual Conference

[41] Qiu, Z. C., Zhang, X. M., Wu, H. X., & Zhang, H. H. (2007). Optimal placement and active vibration control for piezoelectric smart flexible cantilever plate. Journal of Sound and Vibration, 301(3), 521-543

[42] Song, G., Qiao, P. Z., Binienda, W. K., & Zou, G. P. (2002). Active vibration damping of composite beam using smart sensors and actuators. Journal of Aerospace Engineering, 15(3), 97-103

[43] Song G, Schmidt S P and Agrawal B N, Experimental robustness study of positive position feedback control for active vibration suppression, J. Guidance, VOL. 25, NO. 1 ENGINEERING NOTES, 2001

[44] Long, Z., & Guangren, D. (2012, May). A robust vibration suppression controller design for space flexible structures. In Control and Decision Conference (CCDC), 2012 24th Chinese (pp. 4020-4024). IEEE

[45] Han, J. H., & Lee, I. (1999). Optimal placement of piezoelectric sensors and actuators for vibration control of a composite plate using genetic algorithms. Smart Materials and Structures, 8(2), 257

[46] Qiu, Z. C., Zhang, X. M., Wu, H. X., & Zhang, H. H. (2007). Optimal placement and active vibration control for piezoelectric smart flexible cantilever plate. Journal of Sound and Vibration, 301(3), 521-543

[47] Moheimani, S. O. R., Vautier, B. J. G., & Bhilkkaji, B. (2005, December). Multivariable PPF control of an active structure. In Decision and Control, 2005 and 2005 European Control Conference. CDC-ECC'05. 44th IEEE Conference on (pp. 6824-6829). IEEE

[48] Orszulik, R. R., & Shan, J. (2012). Active vibration control using genetic algorithm-based system identification and positive position feedback. Smart Materials and Structures, 21(5), 055002

[49] Kwak, M. K., & Han, S. B. (1998, July). Application of genetic algorithms to the determination of multiple positive-position feedback controller gains for smart structures. In 5th Annual International Symposium on Smart Structures and Materials (pp. 637-648). International Society for Optics and Photonics

[50] Kwak, M. K., & Shin, T. S. (1999, June). Real-time automatic tuning of vibration controllers for smart structures by genetic algorithm. In Proceedings of SPIE- The International Society for Optical Engineering (Vol. 3667, pp. 679-690)

[51] Kwak, M. K., & Heo, S. (2000, June). Real-time multiple-parameter tuning of PPF controllers for smart structures by genetic algorithms. In SPIE's 7th Annual International Symposium on Smart Structures and Materials (pp. 279-290). International Society for Optics and Photonics

[52] Kwak, M. K., & Heo, S. (2007). Active vibration control of smart grid structure by multiinput and multioutput positive position feedback controller. Journal of Sound and Vibration, 304(1), 230-245

[53] Han, J. H., & Lee, I. (1999). Optimal placement of piezoelectric sensors and actuators for vibration control of a composite plate using genetic algorithms. Smart Materials and Structures, 8(2), 257

[54] Kwak, M. K., Heo, S., & Jeong, M. (2009). Dynamic modelling and active vibration controller design for a cylindrical shell equipped with piezoelectric sensors and actuators. Journal of Sound and Vibration, 321(3), 510-524

[55] Ojeda, X., Mininger, X., Gabsi, M., & Lécrivain, M. (2008). Noise cancellation of 6/4 switched reluctance machine by piezoelectric actuators: Optimal design and placement using genetic algorithm

[56] Qiu, Z. C., Wu, H. X., & Ye, C. D. (2009). Acceleration sensors based modal identification and active vibration control of flexible smart cantilever plate. Aerospace Science and Technology, 13(6), 277-290

[57] Qiu, Z. C., Zhang, X. M., Wu, H. X., & Zhang, H. H. (2007). Optimal placement and active vibration control for piezoelectric smart flexible cantilever plate. Journal of Sound and Vibration, 301(3), 521-543

[58] S.S. Groothuis and R.J. Roesthuis, February 2010. Vibration Cancellation Using Synthetic Shunt Impendances. Flinders University, Adelaide, Australia

[59] Moheimani, S. R., Halim, D., & Fleming, A. J. (2003). Spatial Control of Vibrations: Theory and Experiments. World scientific

[60] Halim, D., & Moheimani, S. O. (2004). Reducing the effect of truncation error in spatial and pointwise models of resonant systems with damping. Mechanical systems and signal processing, 18(2), 291-315

[61] S. O. R. Moheimani. Experimental verification of the corrected transfer function of a piezoelectric laminate beam. IEEE Transactions on Control Systems Technology, 8(4):660-666, July 2000

[62] Omidi, E., & Mahmoodi, S. N. (2014). Vibration control of collocated smart structures using H∞ modified positive position and

velocity feedback. Journal of Vibration and Control, 1077546314548471.

[63] Ferrari, G., & Amabili, M. (2015). Active vibration control of a sandwich plate by non-collocated positive position feedback. Journal of Sound and Vibration, 342, 44-56.

[64] Omidi, E., Mahmoodi, S. N., & Shepard, W. S. (2015). Vibration reduction in aerospace structures via an optimized modified positive velocity feedback control. Aerospace Science and Technology, 45, 408-415.

YOUR KNOWLEDGE HAS VALUE

- We will publish your bachelor's and master's thesis, essays and papers

- Your own eBook and book - sold worldwide in all relevant shops

- Earn money with each sale

Upload your text at www.GRIN.com and publish for free